WATCH THEM GROW!

The Life Cycle of a BEETLE

Elaine McKinnon

PowerKiDS press.
New York

Published in 2016 by The Rosen Publishing Group, Inc.
29 East 21st Street, New York, NY 10010

First Edition

Editor: Caitie McAneney
Book Design: Reann Nye

Photo Credits: Cover Anne Sorbes/Moment Select/Getty Images; p. 5 (inset) Serg64/Shutterstock.com; p. 5 (main) Worraket/Shutterstock.com; p. 7 vblinov/Shutterstock.com; p. 9 Marek R. Swadzba/Shutterstock.com; p. 10, 23 (eggs) noppharat/Shutterstock.com; pp. 13, 23 (larva, ladybug), 24 PHOTO FUN/Shutterstock.com; p. 15 Anneka/Shutterstock.com; pp. 16, 23 (pupa), 24 © iStockphoto.com/Henrik_L; pp. 19, 24 Henrik Larsson/Shutterstock.com; p. 20 © iStockphoto.com/Alcuin.

Library of Congress Cataloging-in-Publication Data

McKinnon, Elaine, author.
The life cycle of a beetle / Elaine McKinnon.
pages cm. — (Watch them grow!)
Includes bibliographical references and index.
ISBN 978-1-4994-0660-3 (pbk.)
ISBN 978-1-4994-0661-0 (6 pack)
ISBN 978-1-4994-0662-7 (library binding)
1. Beetles—Juvenile literature. 2. Beetles—Life cycles—Juvenile literature. I. Title. II. Series: Watch them grow!
QL576.2.M35 2015
595.76—dc23

2014048534

Manufactured in the United States of America

CPSIA Compliance Information: Batch #WS15PK: For Further Information contact Rosen Publishing, New York, New York at 1-800-237-9932

Contents

There are many kinds of beetles.
They live all around the world.

A beetle's body changes as it grows. These changes make up its life cycle.

A beetle starts out in an egg. A mother beetle lays hundreds of eggs at a time.

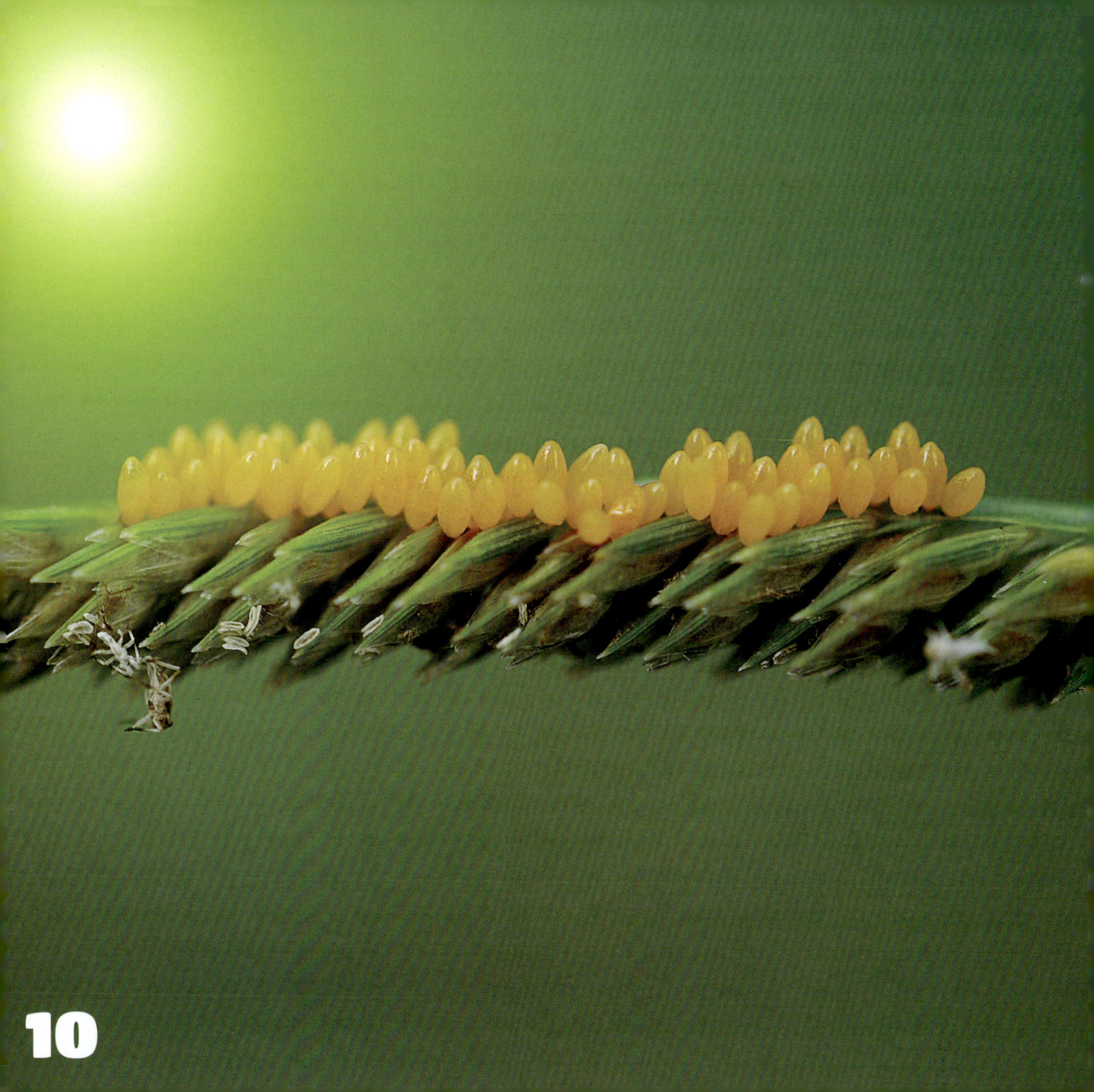

Next, the beetle breaks out of its egg. That's called hatching.

A baby beetle is called a **larva**. It doesn't have wings. The larva eats a lot!

A larva grows fast. It sheds its skin as it grows.

Then, the larva becomes a **pupa**. Pupas do not eat or move. They just grow.

Then, the pupa turns into an adult. Now, it has a hard **shell**.

shell

An adult beetle has four wings and six legs.

Adult beetles can lay eggs. That starts the life cycle all over again!

Life Cycle of a Beetle

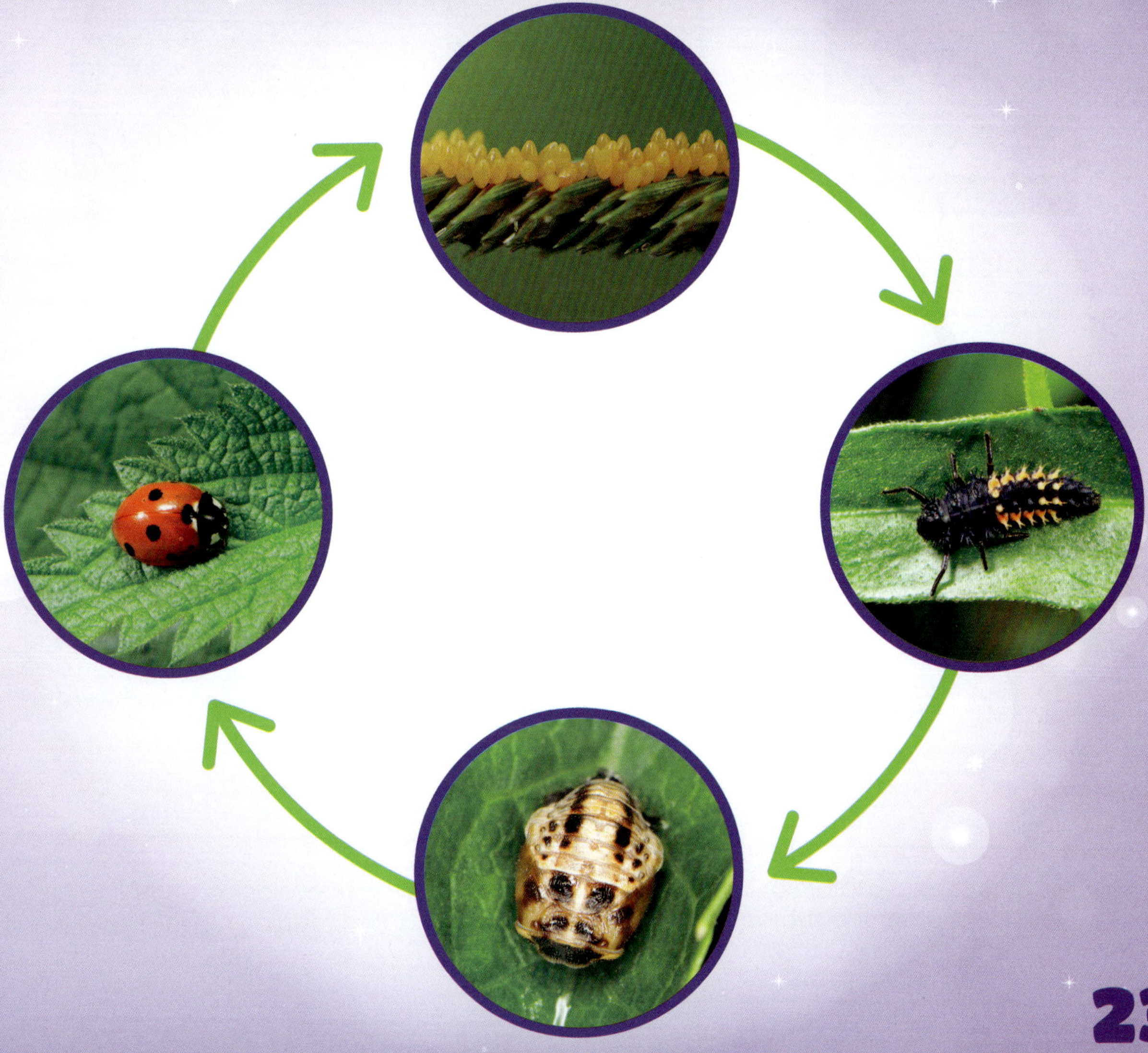

Words to Know

larva

pupa

shell

Index

Websites

Due to the changing nature of Internet links, PowerKids Press has developed an online list of websites related to the subject of this book. This site is updated regularly. Please use this link to access the list: www.powerkidslinks.com/wtg/beet